Workbook to accompany

S a u n d e r s

Core Concepts
in Physics
CD-ROM

Workbook by
Brooke Pridmore, Ph.D.
Clayton College & State University

with problems by
Ray Serway, Ph.D.

For technical support, call 1-800-447-9457.
from 7am to 6pm CST
http://www.hbtechsupport.com

Text typeface: Berkeley Medium

Cover and text design: Jennifer Dunn

Production:
Alexandra Albin
Doris Bruey
Alexandra Buczek
Kate Davey
Paul Graham
Pat Harman
Bruce Hoffman
Samson Jarso
Sherrill Meaney
Diane Southworth
York Production Services

Printed in the United States of America
ISBN 0-03-020038-5
789 021 987654321

CONTENTS

INTRODUCTION

Saunders Core Concepts in Physics for Macintosh® and Windows™ is an interactive, three-disc CD-ROM presentation of introductory, calculus-based physics for college and university students.

FEATURES OF
THE CD-ROM

The presentation:

- Uses live video, animation, interactive graphics, audio, and text to teach fundamental principles of introductory physics.
- Applies the presented concepts to real world phenomena.
- Bridges physical principles to the mathematics that describe them.
- Provides tools for learning and doing physics.

Our goal is to help you develop a deep and practical understanding of physical phenomena, to directly assist you in your study of physics. Because problem-solving is such an essential skill for success in physics, we have also included worked problems and "pop questions" within the presentation.

This three-disc set contains 14 modules (similar to chapters in a textbook) and is accompanied by this workbook, which can be used in addition to any other general physics text.

These discs also include:

- Tools such as a Unit Converter and Physical Constants table
- Notebook function
- Active indexes of contents and equations
- Worked Sample Problems and Pop Questions

INTERACTIVE PRESENTATION

Information is presented as "screens" within modules. Each screen in a module introduces a key concept or a set of related concepts. Complete instructions for using this presentation are in the section of this workbook entitled *Using the CD-ROM*, and in the instructions that accompany the discs.

WORKBOOK

The workbook is organized around the main concept screens, each of which is numbered by the module and screen. The questions in the workbook can be answered after reading the text on the screen (and related screens), viewing the media, and working through associated problems.

USER'S GUIDE

SYSTEM
REQUIREMENTS

To use the applications included on the CD, your system must meet the following requirements:

Macintosh installation requires:

- Macintosh computer running System 7.0.1 or greater
- 4 megabytes of free RAM minimum after system loads (5 megabytes preferred)
- 2 megabytes available hard disk space
- 256 color display at 640 × 480 pixels (minimum)
- Double-speed CD-ROM drive (minimum)

Windows installation requires:

- 486DX-33 computer running Windows 3.1 or greater
- 8 megabytes of RAM or greater
- 256 color capable VGA video board that is MPC Level II compliant, and a color VGA display (minimum)
- SoundBlaster™ compatible audio card, and speakers or a headset
- 3 megabytes available hard disk space
- Double-speed CD-ROM drive (minimum)

Installation of the *Saunders Core Concepts in Physics* presentation can be performed from any disc in the three-disc set. Only one installation is needed. Each disc contains both Macintosh and Windows versions of the presentation.

Installation Procedure for Macintosh

> *Note: Installation is necessary only if your system does not already have QuickTime™ version 2.0 or later. If it does, no installation is necessary.* Saunders Core Concepts in Physics *installs version 2.5 of QuickTime. (QuickTime is Apple Computer's programming "architecture" for motion pictures.)*

Put the disc in the CD-ROM drive and double-click the *Core Concepts* disc icon on the desktop to open the CD-ROM window.

In the CD-ROM window, double-click the QuickTime Installer icon.

The installation window appears with the "Standard Installation" option preselected.

> *Note: If you are using System 7.5.0 or an earlier version, you will need Sound Manager.*
> *To install Sound Manager, press and hold down the shift key while selecting "Sound Manager."*

Click the Install button. The window indicating that the installation was successful appears. Select "Quit." (QuickTime will be placed in the Extensions folder, which is found in your System folder.)

Installation is now complete. You may need to restart your computer to run the presentation.

> *Note: To run the* Saunders Core Concepts in Physics *presentation, one of the discs must be in the CD-ROM drive. Double-click on the Physics icon.*

Installation Procedure for Windows

> *Note: Installation is necessary only if your system does not already have QuickTime™ version 2.03 (or later) for Windows. If it does, no installation is necessary. Saunders Core Concepts in Physics installs version 2.1.2 of QuickTime for Windows. (QuickTime is Apple Computer's programming "architecture" for motion pictures.)*

Put the disc in the CD-ROM drive.

Windows '95 and NT 4.0 Systems
Click on the "Start" button in the left corner of the Windows 95 taskbar.

Select "Run" from the pop-up menu. Type **D:\QT16.EXE** and choose "OK."

*Note: If your CD-ROM drive is not accessed through the drive letter "D,"
substitute the appropriate letter in the setup command.*

After reading the Software License Agreement, select "Agree" to continue the installation. Select "Install" from the Begin Install dialogue box.

From the Check Existing Versions dialogue box, select "Start." To complete installation, select "Install" from the Complete Install dialogue box. (QuickTime files will be placed in the System directory, which is found in your Windows directory.)

Once files have been installed successfully, select "Play Sample" for sample movie. Then select "Exit" from the File menu to exit the Movie Player.

Installation is now complete.

Note: To run the Saunders Core Concepts in Physics *presentation, one of the
discs must be in the CD-ROM drive. Double-click on the Physics icon.*

Windows 3.1 or NT 3.5 Systems

From Program Manager, select "Run" from the file menu. Type **D:\QT16.EXE** and choose "OK."

*Note: If your CD-ROM drive is not accessed through the drive letter "D,"
substitute the appropriate letter in the setup command.*

After reading the Software License Agreement, select "Agree" to continue the installation. Select "Install" from the Begin Install window.

From the Check Existing Versions window, select "Start." To complete installation, select Install from the Complete Install window. (QuickTime files will be placed in the System directory, which is found in your Windows directory.)

Once files have been installed successfully, select "Play Sample" for sample movie. Then select "Exit" from the file menu to exit the Movie Player.

Installation is now complete.

Note: To run the Saunders Core Concepts in Physics *presentation, one of the
discs must be in the CD-ROM drive. Double-click on the Physics icon.*

The *Saunders Core Concepts in Physics* CD-ROM is a complete multimedia presentation of college-level calculus-based introductory physics.

Core Concepts Disc 1

Starting the Presentation for Macintosh
Once the program is properly installed and the desired disc is in the CD-ROM drive, open the CD-ROM by double-clicking on its icon.

To begin the presentation, double-click on the *Physics* icon.

Physics

Starting the Presentation for Windows 95 or NT 4.0 Systems
Once the program is properly installed and the desired disc is in the CD-ROM drive, open "My Computer" and double-click on the CD-ROM icon.

To begin the presentation, double-click on the *Physics.exe* icon.

Physics.exe

Starting the Presentation for Windows 3.1 or NT 3.5 Systems
From Program Manager, select "Run" from the file menu. Type **D:\PHYSICS.EXE** and choose "OK."

> *Note: If your CD-ROM drive is not accessed through the drive letter "D,"*
> *substitute the appropriate letter in the setup command.*

Using the Presentation
A title screen appears with a "Production Credits" bar at the screen's lower right corner. To view credits, move hand cursor and click on the bar. To view the *Contents* screen, click anywhere else on the title screen or credits screen.

The mouse is used for all navigation. Navigation within the presentation is accomplished by a single click of the mouse.

A pointing finger cursor 🖑 indicates an active area.

Inactive screen areas are indicated by the arrow cursor 🠔 .

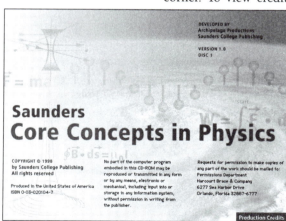

[Title Screen]

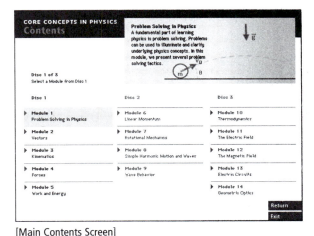

[Main Contents Screen]

The presentation is divided into modules, with Modules 1–5 on Disc 1, Modules 6–9 on Disc 2, and Modules 10–14 on Disc 3.

Click on the desired module to launch it from the *Contents* screen.

In the Contents, all active material is highlighted; all inactive material is dimmed. The material from discs other than the one in use can be accessed only by inserting the desired disc, although synopses for all modules are viewable on each disc by rolling the mouse over the module title.

The first screen of every module is a *Contents and Introduction* screen. From this screen, view the introductory "touchstone" movie or click on a topic name to open the desired Main Screen.

Each module is organized into a series of Main Screens, which address a single topic or a group of closely related topics. The *Contents and Introduction* screen provides a list of that module's Main Screens, as well as the introductory movie.

[Contents and Introduction Screen]

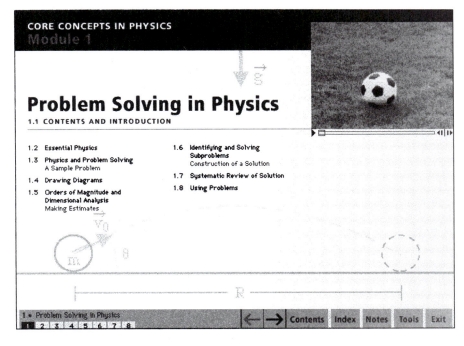

Navigation Bar Operation

At the bottom of the screen is the navigation bar.

The number of the screen currently in use is highlighted.

In the navigation bar, the arrow pointing left (the Back arrow) allows the user to return to the previous Main Screen. The arrow pointing right (the Forward arrow) takes the user to the next Main Screen.

The Index button accesses the index of important terms and concepts used throughout the presentation.

Selecting the Tools button displays a pop-up list of interactive tools.

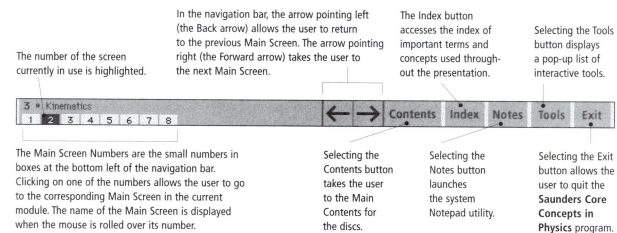

The Main Screen Numbers are the small numbers in boxes at the bottom left of the navigation bar. Clicking on one of the numbers allows the user to go to the corresponding Main Screen in the current module. The name of the Main Screen is displayed when the mouse is rolled over its number.

Selecting the Contents button takes the user to the Main Contents for the discs.

Selecting the Notes button launches the system Notepad utility.

Selecting the Exit button allows the user to quit the **Saunders Core Concepts in Physics** program.

Main Screens

Main Screens are accessed either from the Table of Contents on the Module *Contents and Introduction* screen or from the Navigation Bar at the bottom of each screen in a Module. Each Main Screen includes several features such as video or audio clips, problems, sidebars, tables, math-in-detail banners, or animated simulations that provide information about the current topic.

Many Main Screens consist of two sections. The second section can be accessed by selecting a colored bar in a corner of the screen.

Most features can be accessed and navigated by clicking on their corresponding colored arrows.

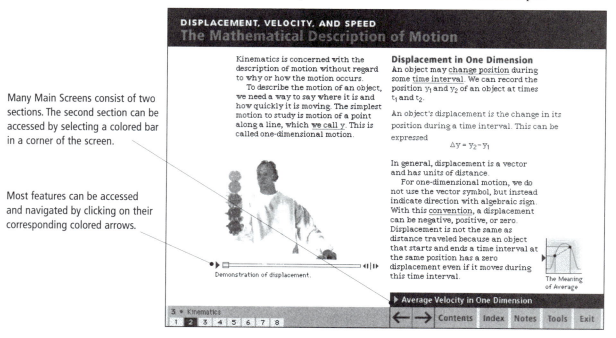

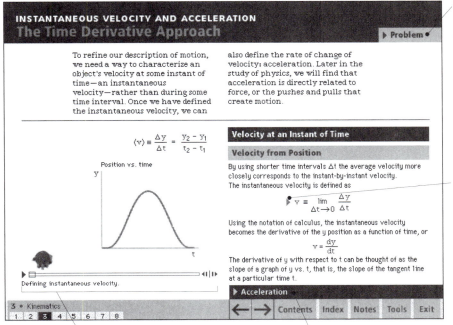

The Problem button is found in the top right corner of some screens. Select it to access a problem related to the current topic.

The arrows in the screen are used to access more information or to initiate an action such as playing an animation, or displaying mathematics-in-detail banners.

At the lower edge of most movies is a sliding control bar with play/pause button, which allows the user to see the movie with narration. The slider can be manipulated using the mouse to move forward or backward through the clip.

A Section Bar is often found at the bottom right of Main Screens; selecting it accesses the next section of the Main Screen.

Pop-up Questions are indicated by the icon on many screens. Select the icon for the question. To view answer choices, select the "A" in the question box. The correct answer appears after you select your answer choice.

Place the cursor over any underlined text to access a definition or explanation of that term.

[Tools button]

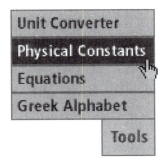

Accessing the Tools Menu

Interactive tools are available from every Presentation Screen, accessed from the Tools button in the Navigation Bar. Selecting and holding the button displays a pop-up list of the tools: the Unit Converter, Equations index, Physical Constants and Physical Data tables, and more. To access a tool, roll over its name and release.

To return to the presentation screen from the tool, click on the return bar at the bottom right of the menu.

user's Guide

Reference Section and Troubleshooting Guide

This section is organized by:

- Navigation Functions
- Media Access Functions
- Utilities

Navigation Functions

[Forward/Backward Arrows]

Forward and reverse arrows allow the user to move to the next or previous Main Screen.

> *Note: The Module Opener contains no active reverse arrow, and no active forward arrow appears on the last screen of a Module.*

[Contents button]

The Contents button allows the user to return to the *Contents* screen. It also allows the user to move from one Module to another on the active disc.

[Main Screen Numbers]

At the bottom of every Module screen is a sequence of numbers corresponding to the Main Screens of that Module. The current screen is highlighted. Moving the mouse over these numbers displays the name of each Main Screen. Clicking on a number takes the user to that Main Screen.

[Exit button]

The Exit button quits the presentation.

> *Note: The* Saunders Core Concepts in Physics *program can also be terminated by choosing Quit from the file menu.*

Media Access Functions

[Colored Arrows]

Small colored arrows, found throughout the *Saunders Core Concepts in Physics* presentation, initiate some action such as accessing mathematics-in-detail banners, sidebars, or other features.

[Video Controller]

II
[Pause]

Video buttons allow the user to play, stop, pause, and replay video or animation clips. Most clips have a sliding controller that allows users to advance or reverse the clip.

> *Note: If the video plays poorly or drops frames, or if the audio tends to "cut-out," make sure you have quit any open applications to free up additional RAM. If the problem persists, try running the application on a computer with more memory or a faster CD-ROM drive.*

Utilities

The Navigation Bar also includes the following utilities.

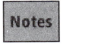

[Notebook access button]

The Notebook utility allows the user to enter, save, and print notes while using the *Saunders Core Concepts in Physics* program. The utility uses the standard Notepad application provided with Apple Macintosh or Microsoft Windows operating systems.

> *Note: Instructions for use of the Notebook utility are available in the documentation for the Notepad function in the operating system user's guide.*

If clicking on the Notes button does nothing, verify that the Notepad utility is installed on the system.

On Macintosh computers, the contents of the Notepad file are saved automatically. The file can be saved and renamed as a separate document, if desired.

In Windows, the first time you access the Notepad you will be prompted to create a file called "Physnote.txt." Choose "Yes." To keep your notes from session to session, you must save them. The file can be saved and renamed as a separate document, if desired.

[Index button]

[Index Screen]

The Index of important terms and concepts allows the user to find any important topic or term featured within the presentation. Select the colored numbers to move to a reference on the current disc.

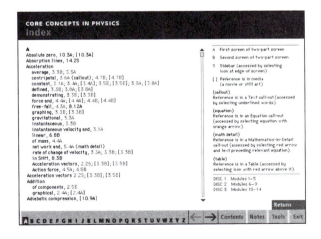

MODULE 1

Problem Solving in Physics

It has been said that practice is the best instructor. In the study of physics, "practice" means solving problems. This workbook is designed to help you get practice.

Before tackling any physics problem, you will find it useful to equip yourself with a set of mental tools. We strongly recommend that you view the *Problem Solving in Physics* module as a first step in acquiring these tools, and to continue to refer back to it as you progress in your studies.

Here is a checklist of reminders for solving most of the problems you are likely to encounter in your physics course:

- Draw a diagram
- Estimate order of magnitude
- Perform dimensional analysis
- Identify and solve subproblems
- Construct a final answer
- Systematically review your answer
  - Check order of magnitude
  - Check units/dimensions
  - Interpret your equations for "sanity"
  - Check special cases

You will very likely wish to adapt and build on these tools as you develop your own strategies and tactics for solving problems. A key point to remember is that learning physics does not demand memorizing innumerable equations; instead, it calls for the building of bridges between various concepts and principles in physics.

MODULE 2   Vectors

Physical phenomena can often be quantized by magnitude alone (e.g., the temperature at a given point in space). Other times, they can be specified by both a direction and a magnitude (e.g., the wind's speed and direction at the same point). Quantities that have both magnitude and direction are described by vectors. In this module, we introduce vectors and some of their applications and operations (such as addition, subtraction, and multiplication). We also discuss polar and cartesian coordinate systems, and why choosing a particular coordinate system can simplify an analysis.

DEFINITIONS

We use the three trigonometric functions:

$$\sin \theta = \frac{\text{opposite}}{\text{hypotenuse}} \qquad \cos \theta = \frac{\text{adjacent}}{\text{hypotenuse}} \qquad \tan \theta = \frac{\text{opposite}}{\text{adjacent}}$$

and the Pythagorean theorem:

$$(\text{hypotenuse})^2 = (\text{adjacent})^2 + (\text{opposite})^2$$

Vector components.

If $\vec{A} + \vec{B} = \vec{C}$, then $A_x + B_x = C_x$ and $A_y + B_y = C_y$

Scalar (dot) product.

$\vec{A} \cdot \vec{B} = |AB| \cos \theta$, where θ = angle between $\vec{A}$ and $\vec{B}$

$\vec{A} \cdot \vec{B} = A_x B_x + A_y B_y + A_z B_z$

Vector (cross) product.

$|\vec{A} \times \vec{B}| = |AB| \sin \theta$, where θ = angle between $\vec{A}$ and $\vec{B}$

Right-hand rule. Orient the right hand such that the fingers curl in the direction of $\vec{A}$ toward $\vec{B}$ through the smallest angle. Lift the thumb of the right hand; it will point in the direction of $\vec{A} \times \vec{B}$.

$$\vec{A} \times \vec{B} = \begin{vmatrix} \hat{i} & \hat{j} & \hat{k} \\ A_x & A_y & A_z \\ B_x & B_y & B_z \end{vmatrix}$$

$$\vec{A} \times \vec{B} = (A_y B_z - A_z B_y)\hat{i} - (A_x B_z - A_z B_x)\hat{j} + (A_x B_y - A_y B_x)\hat{k}$$

Coordinate Systems

Problem Description

Two points in a plane have polar coordinates $(r, \theta) = (2.50 \text{ m}, 30.0°)$ and $(3.80 \text{ m}, 120.0°)$, respectively. Determine the cartesian coordinates of these points and the distance between them.

Before we begin...

1. Draw a diagram indicating the two vectors as $\vec{A}$ and $\vec{B}$.

2. How are the cartesian coordinates x and y related to the polar coordinates, r and θ?

Solving the problem

3. For each vector, $\vec{A}$ and $\vec{B}$, find the x and y components.

4. The vector separating the two points is $\vec{B} - \vec{A}$. Find the x and y coordinates of $\vec{B} - \vec{A}$.

5. Use the Pythagorean theorem to find the distance, which is the magnitude of $\vec{B} - \vec{A}$.

Vector Addition and Subtraction

Problem Description

A force $\vec{F}_1$ of magnitude 6.00 units acts on an object at the origin in a direction 30.0° above the positive x axis. A second force $\vec{F}_2$ of magnitude 5.00 units acts on the same object in the direction of the positive y axis. Use a graph to determine the magnitude and direction of the resultant force $\vec{F}_1 + \vec{F}_2$.

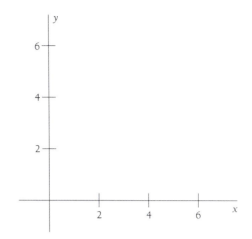

Before we begin...   1. Draw the forces from the origin in the diagram shown here.

Solving the problem   2. Given the forces as drawn, use graphical addition to find the resultant force.

Vector Components and Unit Vectors

Problem Description

A displacement vector lying in the *xy* plane has a magnitude of 50.0 m and is directed at an angle of 120.0° above the positive *x* axis. What are the rectangular components of this vector?

Before we begin...

1. The vector can be expressed in polar coordinate form as $\vec{A}$ = (50.0 m, 120.0°). How are the rectangular coordinates of a vector related to the polar coordinates?

2. Draw the vector on the above axis system. Drop perpendicular lines from the tip of the vector to the *x* axis and to the *y* axis. These projections give the length of A_x and A_y, respectively.

Solving the problem

3. Use the information in the above answer to find the rectangular coordinates A_x and A_y.

Vector Components and Unit Vectors

Problem Description

Instructions for finding a buried treasure include the following: Go 75 paces at 240°, turn to 135° and walk 125 paces, then travel 100 paces at 160°. Determine the resultant displacement from the starting point.

Before we begin...

1. Draw a sketch of the problem, using the graphical method of adding vectors.

2. Express the three displacement vectors $\vec{A}$, $\vec{B}$, and $\vec{C}$ in polar coordinates.

   $\vec{A} =$ $\vec{B} =$ $\vec{C} =$

Solving the problem

3. Find the x and y components of the three vectors.

(continued on next page . . .)

4. Add the x components together to get the x component of the total displacement. Do the same for the y components.

5. Use the Pythagorean theorem to find the magnitude of the resultant vector.

6. Use a suitable trigonometric function to determine the angle that the resultant vector makes with respect to the x axis.

The Dot Product

Problem Description

Vector $\vec{A}$ extends from the origin to a point having polar coordinates (7, 70°) and vector $\vec{B}$ extends from the origin to a point having polar coordinates (4, 130°). Find $\vec{A} \cdot \vec{B}$.

Before we begin...

1. Sketch the vectors $\vec{A}$ and $\vec{B}$ on the coordinate system above. Indicate the angle that is formed between the two vectors.

2. What are the magnitudes of the two vectors?

   $|\vec{A}| =$ $|\vec{B}| =$

3. Is $\vec{A} \cdot \vec{B}$ a vector or a scalar quantity?

Solving the problem

$\vec{A} \cdot \vec{B}$ is the scalar product of the two vectors. In this problem, the vectors are expressed in polar coordinates. The scalar product is calculated according to the relation:

$$\vec{A} \cdot \vec{B} = |\vec{A}||\vec{B}| \cos \theta$$

where θ is the angle between $\vec{A}$ and $\vec{B}$.

4. Use the information gathered above to compute the value.

The Dot Product

Problem Description
Vector $\vec{A}$ is 2.0 units long and points in the positive y direction. Vector $\vec{B}$ has a negative x component 5.0 units long, a positive y component 3.0 units long, and no z component. Find $\vec{A} \cdot \vec{B}$ and the angle between the vectors.

Before we begin...

1. In this problem, the vectors are expressed in a different form than in the previous example. What form does this problem use?

2. Identify the x, y and z components of each vector:

$A_x =$ $A_y =$ $A_z =$

$B_x =$ $B_y =$ $B_z =$

Solving the problem

3. The scalar product in rectilinear coordinates (x, y, z) is computed according to the relationship:

$$\vec{A} \cdot \vec{B} = A_x B_x + A_y B_y + A_z B_z$$

Compute the scalar product.

Notice that the scalar product is a quantity without direction. It is not a vector. If you know the scalar product, you can use the relationship:

$$\vec{A} \cdot \vec{B} = |A||B| \cos \theta$$

to find θ, the angle between the two. In order to find θ, what do you need to compute before substituting into the relationship?

4. Calculate the angle θ.

The Cross Product

Problem Description
Given $\vec{M} = 6\hat{i} + 2\hat{j} - \hat{k}$ and
$\vec{N} = 2\hat{i} - \hat{j} - 3\hat{k}$, calculate $\vec{M} \times \vec{N}$.

Before we begin...

1. Identify the components of the two vectors:

   $M_x =$ $M_y =$ $M_z =$

   $N_x =$ $N_y =$ $N_z =$

2. Is the resultant of this multiplication going to yield a vector or a scalar quantity?

Solving the problem

3. Because we have the (x, y, z) components of the two vectors, the relationship resulting from evaluating the determinant:

$$\vec{M} \times \vec{N} = \begin{vmatrix} \hat{i} & \hat{j} & \hat{k} \\ M_x & M_y & M_z \\ N_x & N_y & N_z \end{vmatrix}$$

$$\vec{M} \times \vec{N} = (M_y N_z - M_z N_y)\hat{i} - (M_x N_z - M_z N_x)\hat{j} + (M_x N_y - M_y N_x)\hat{k}$$

is used to compute $\vec{M} \times \vec{N}$.

The Cross Product

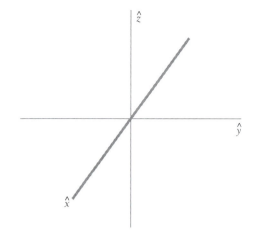

Problem Description

Vector $\vec{A}$ is in the negative *y* direction and vector $\vec{B}$ is in the negative *x* direction. What is the direction of $\vec{A} \times \vec{B}$? What is the direction of $\vec{B} \times \vec{A}$?

Before we begin...

1. Sketch the vectors on the above coordinate system. Do not worry about the lengths—only the directions are important. Why?

2. What is the relationship between the vector product of two vectors and the plane formed by the two vectors?

Solving the problem

3. Using the definition of the vector product

$$\vec{A} \times \vec{B} = |A|\,|B|\,\sin\theta$$

where θ is the angle between $\vec{A}$ and $\vec{B}$, determine the value of θ as measured counterclockwise (as per the right-hand rule).

4. If the angle is 90°, sin θ is +1; if the angle is 270°, sin θ is −1. What does this tell you about the direction of $\vec{A} \times \vec{B}$?

In this module, we describe and quantify the motion of objects in one and two dimensions. This is a necessary first step before we can hope to learn about the causes of motion, which we begin to explore in Module 4, *Forces*.

The concepts used to describe motion, such as displacement, velocity, and acceleration, are vectors. For motion in one direction (i.e., back and forth along a straight-line path), a simple plus or minus sign indicates the direction of these vectors. For two- and three-dimensional motion, we use the full set of vector operations we studied in Module 2, *Vectors*.

Displacement. The change in position of an object represented by Δx in one-dimensional motion and $\Delta \vec{r}$ in more than one dimension. It is a vector quantity that is computed by

$$\Delta \vec{r} = \vec{r}_f - \vec{r}_i$$

Velocity, $\vec{v}$. The rate of change of displacement with respect to time. The instantaneous velocity can be computed by

$$\vec{v} = \frac{d\vec{r}}{dt}$$

On a graph of displacement as a function of time, the slope of the tangent line to the displacement at the time gives the instantaneous velocity.

Acceleration, $\vec{a}$. The rate of change of velocity with respect to time.

$$\vec{a} = \frac{d\vec{v}}{dt} = \frac{d^2\vec{r}}{dt^2}$$

(*Note:* In one dimension we use x or y rather than r for the position coordinate.)

$$\vec{r} = r_x \hat{i} + r_y \hat{j}$$

For constant acceleration

$$\vec{v} = \vec{v}_0 + \vec{a}t$$

$$\vec{r} = \vec{r}_0 + \vec{v}_0 t + \tfrac{1}{2}\vec{a}t^2$$

$$\vec{v}^2 = \vec{v}_0{}^2 + 2\vec{a} \cdot \vec{r}$$

For uniform circular motion

$$\vec{a}_c = v^2/r$$

toward the center of the circle.

For nonuniform circular motion

$$\vec{a} = \vec{a}_t + \vec{a}_r$$

where $\vec{a}_t$ is the tangential component of the acceleration.

Displacement, Velocity, and Speed

Problem Description

The velocity of a particle as a function of time is shown. At $t = 0$, the particle is at $x = 0$. Sketch the acceleration as a function of time. Determine the average acceleration of the particle from time $t = 2.0$ s to $t = 8.0$ s. Determine the instantaneous acceleration of the particle at $t = 4.0$ s.

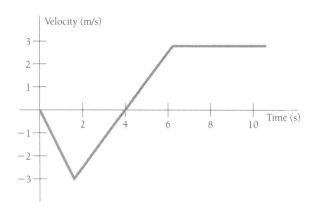

Before we begin...

The graph of velocity as a function of time given in this problem consists of three straight line segments, from (0 s, 0 m/s) to (2 s, −3 m/s) to (6 s, 3 m/s) to (8 s, 3 m/s).

1. How is the acceleration vector related to velocity vs. time?

Solving the problem

Because the average acceleration is equal to the rate of change of the velocity, we can use the relation

$$\langle \vec{a} \rangle = \Delta \vec{v}/\Delta t$$

to find the average acceleration over any time interval.

2. Does the acceleration change in the time between $t = 0$ and $t = 2$ s? What is the value of the acceleration?

3. Does the acceleration change in the time between $t = 2$ and $t = 6$ s? What is the value of the acceleration?

(continued on next page . . .)

4. Does the acceleration change in the time between $t = 6$ and $t = 8$ s?

What is the value of the acceleration?

5. You can now graph acceleration vs. time, using the answers to the above questions.

6. The average acceleration between any two time values is given by $\langle \vec{a} \rangle = \Delta \vec{v} / \Delta t$. Use the data from the original graph to answer questions relating to average acceleration.

7. Use the slope of the tangent line to the velocity vs. time graph to evaluate the acceleration at a particular point in time.

Instantaneous Velocity and Acceleration

Problem Description
The position of a softball tossed vertically
upward is described by the equation
$y = 7.00t - 4.90t^2$, where y is in meters
and t in seconds. Find the ball's initial
speed v_0 at $t = 0$. What is its velocity at
$t = 1.26$ s? What is its acceleration?

Before we begin...

1. How are instantaneous velocity and acceleration related to displacement as a function of time?

Solving the problem

2. To find the initial speed of the ball at $t = 0$, we first find the velocity as a function of time. This will then be evaluated by substituting $t = 0$ into the equation. The resulting magnitude is the speed.

3. Evaluate the velocity at $t = 1.26$ s. In this portion of the problem, the sign of the answer is important. It indicates the direction of the velocity.

4. Evaluate the acceleration, again using calculus.

One-Dimensional Motion at
Constant Acceleration

Problem Description
A hockey player is standing on a frozen
pond when an opposing player skates by
with the puck, moving with a uniform
speed of 12.0 m/s. After 3.00 s, the first
player makes up his mind to chase his opponent. If the first player accelerates uni-
formly at 4.00 m/s², how long does it take
him to catch his opponent? How far has
the first player traveled during this time?

Before we begin...

This problem involves describing the motion of two different objects (players). One player, called the first, undergoes an acceleration while the second
player moves at constant speed.

1. When the first player catches the second, what can be said about the distance that each player has traveled? (In particular, who will have traveled
   farther?)

2. Identify the given information concerning the motion of each player:

First Player	Second Player
$v_{01} =$	$v_{02} =$
$a_1 =$	$a_2 =$

3. When the first player catches the second, how much time, t_2, has the
   second player been moving? (Assume the first player has been moving
   t_1 seconds.)

SHM and Waves in the Real World

Problem Description
A simple pendulum has a length of 3.00 m. Determine the change in its period if it is taken from a point where $g = 9.80$ m/s^2 to an elevation where the free-fall acceleration decreases to 9.79 m/s^2.

Before we begin...

1. For small amplitudes, what is the relation between the period of motion of a simple pendulum, the pendulum's length, and the acceleration due to gravity?

2. How do you calculate the change in a quantity as measured under two different conditions?

Solving the problem

3. Use the equation relating the period of motion to the length of a simple pendulum and the acceleration due to gravity to calculate the periods for each of the given acceleration values.

4. Subtract the value computed for $g = 9.80$ m/s^2 from the value computed when $g = 9.79$ m/s^2.

MODULE 9

Wave Behavior

In Module 8, *Simple Harmonic Motion and Waves*, we introduced wave motion for traveling waves and defined the concepts of wavelength, frequency, and wave speed. In this module, we investigate how boundary conditions and other waves affect the overall propogation of a wave through a medium, as well as how the energy and power of that wave is affected by such conditions.

Boundary. The interface between any two mediums. Waves at a boundary can be reflected and/or transmitted.

Superposition principle. When two waves combine, they pass straight through without interruption or distortion. At a position y, the total disturbance is the sum of the individual disturbances:

$$y = y_1 + y_2$$

Interference. The addition or subtraction of the amplitudes of two waves located at the same position and time yields an amplitude that is constructive if the two are in phase and destructive if the two are 180° out of phase.

Standing wave. A wave in a confined region that "oscillates in place" rather than moving across space. Mathematically, these wave patterns always result from interference between two or more moving waves.

Harmonics. Stable modes of vibration corresponding to particular frequencies. The lowest allowable frequency is called the first harmonic.

Node. A point of zero amplitude in a standing wave, created by ongoing destructive interference.

Antinode. A position of relative maximum amplitude in a standing wave created by ongoing constructive interference.

Resonance. When energy is added to a system at the system's natural frequency, the amplitude of oscillations is maximized.

USEFUL EQUATIONS

The speed of a one-dimensional mechanical wave such as a pulse in a stretched string is computed by

$$v = \sqrt{\frac{F_t}{\mu}}$$

where F_t is the tension in the string and μ is the linear mass density of the string.

The energy E and the power P associated with a wave are calculated by the equations

$$E = \tfrac{1}{2}\mu\omega^2A^2L \qquad \text{and} \qquad P = \tfrac{1}{2}\mu\omega^2A^2v$$

where ω is the angular frequency of the wave, A is the wave amplitude, L is the length of the string, and v is the speed of the wave.

For standing waves, the relationships between the allowable frequencies of vibration and the length of the string are given by

$$f_n = nv/2L \qquad n = 1, 2, 3...$$ (wave fixed at both ends or free at both ends)

$$f_n = nv/4L \qquad n = 1, 3, 5...$$ (wave fixed at one end and free at the other end)

Speed of a Wave in a Medium

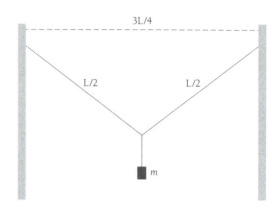

Problem Description

A light string of mass per unit length 8.00 g/m has its ends tied to two walls separated by a distance equal to 3/4 the length L of the string. A mass m is suspended from the center of the string, putting a tension in the string. Find an expression for the transverse wave speed in the string as a function of the hanging mass. How much mass should be suspended from the string to have a wave speed of 60.0 m/s?

Before we begin...

1. Identify the given information:

   $\mu =$ $v =$

2. Draw the free-body diagram for the system.

3. The tension in each section of the string is the same. This ensures that the x components of the forces sum to zero. What do the y components of the tensions have to equal to ensure equilibrium?

4. How can the angle θ that the string makes with respect to the horizontal be computed?

Solving the problem

5. The string and the mass are in equilibrium, requiring that the vector sum of the forces be zero. By symmetry, the x components of the tension will be equal in magnitude to each other. Apply the condition of equilibrium to the y components and solve for the tension T.

6. The relation between the transverse wave speed and the string through which it is traveling is

$$v = \sqrt{\frac{F_t}{\mu}}$$

where F_t is the tension and μ is the linear mass density.

The tension depends upon the weight that is supported. Solve the above given equation for tension and then substitute into the relation between tension and weight.

Energy and Power in Waves

Problem Description

Transverse waves are being generated on a rope under constant tension. By what factor is the required generating power increased or decreased if the length of the rope is doubled and the angular frequency remains constant? The power is changed by what factor when the amplitude is doubled and the angular frequency is halved?

When both the wavelength and amplitude are doubled, what happens to the power? What happens to the power when both the length of the rope and the wavelength are halved?

Before we begin...

1. Write the expression for the power delivered by a transverse wave.

2. Does the length of the rope affect the power?

Solving the problem

3. If the length changes, it does not affect the power. Isolate the dependency for the various combinations of variables described in the four questions asked in the description. Answer the questions.

Superposition and Interference

Problem Description
Two waves are traveling in the same direction along a stretched string. Each has an amplitude of 4.0 cm, and they are 90° out of phase. Find the amplitude of the resultant wave.

Before we begin...

1. Write the wave function of each of the two waves.

2. State the superposition principle.

Solving the problem

3. Recall the trigonometric identity

$$\sin (a) + \sin (b) = 2 \sin ((a + b)/2) \cos (a - b)/2)$$

Let $a = (kx - \omega t)$ and $b = (kx - \omega t - \phi)$. In this problem, $\phi = 90°$. Substitute and simplify the equation.

Standing Waves

Problem Description

Two waves given by $y_1(x, t) = A \sin (kx - \omega t)$ and $y_2(x, t) = A \sin (2kx + \omega t)$ interfere. Determine all x values where there are stationary nodes. Determine all x values where there are nodes that depend on time t.

Before we begin...

1. What are node positions?

2. According to the superposition principle, what must be true for a point to be a node?

Solving the problem

3. Use the superposition principle to find the wave function of the combined waves.

4. Stationary nodes occur as a result of a term that is independent of time t equaling zero. Identify the term in the combined equation, set it equal to zero, and solve for the values of x that satisfy the equation.

5. Time dependent nodes arise from a term that has both position x and time t dependency. Identify such a term in the equation and solve it by setting it equal to zero.

Standing Waves—Wave Fixed at Both Ends

Problem Description
A 2.0-m long wire having a mass of 0.10 kg is fixed at both ends. The tension in the wire is maintained at 20 N. What are the frequencies of the first three allowed modes of vibration?

Before we begin...

1. What is the relationship for the allowable wavelengths for a wave fixed at both ends?

2. How can the speed of the wave be calculated for a wire under tension?

Solving the problem

3. Compute the mass density μ for the wire and use this result to find the speed of the wave.

4. Calculate the frequencies by equating the frequency f_n from the speed and the allowable wavelengths λ_n.

Standing Waves—Wave with One Fixed End and One Free End

Problem Description

A student uses an audio oscillator of adjustable frequency to measure the depth of a water well. Two successive resonant frequencies are heard at 52.0 Hz and 60.0 Hz. What is the depth of the well?

Before we begin...

1. The water level in the well is treated as a node, and the top of the well is an antinode. In this problem, one end is fixed and one end is free. What is the relationship between the length of the system and allowable frequencies for this type of system?

2. Identify the given information:

   $f_a =$ $f_b =$

Solving the problem

3. Write the expression for the allowable frequency with $a = n$ and for the allowable frequency with $b =$ next allowable frequency.

4. Solve the equations simultaneously by setting the difference in frequencies $f_b - f_a$ equal to the difference in their computational formulas. This will eliminate n from the equation. We now have change in frequency as a function of wavespeed and depth of the well.

5. Use the speed $v = 344$ m/s (speed of sound at standard temperature and pressure). Solve for the depth of the well.

Resonance

Problem Description
A weight of 40.0 N is suspended from a spring that has a force constant of 200 N/n. The system is undamped and is subjected to a harmonic force of frequency 10.0 Hz, resulting in a forced-motion amplitude of 2.00 cm. Determine the maximum value of the force.

Before we begin...

1. Identify the given information:

$$m = w/g = \qquad k = \qquad A = \qquad \omega =$$

Let the harmonic force be represented by

$$F = F_0 \sin \omega t = ma_0 \sin \omega t$$

The equation of motion for the system subjected to this driving force is found by applying Newton's second law:

$$m\frac{d^2x}{dt^2} = -kx + ma_0 \sin \omega t$$

Dividing all terms by the mass m gives

$$\frac{d^2x}{dt^2} = -\frac{k}{m}x + a_0 \sin \omega t$$

Upon substituting $\omega_0^2 = k/m$, the equation of motion can be written as

$$\frac{d^2x}{dt^2} = -\omega_0^2 x + a_0 \sin \omega t$$

Solving the problem   Use the trial solution

$$x = C \sin \omega_0 t + A \sin \omega t$$

2. Evaluate the second derivative of x with respect to time t.

3. Substitute the second derivative and the trial solution value of x into the equation of motion.

4. Simplify the equation and solve for A that satisfies it.

5. The maximum value of the force occurs when all of the amplitude of the motion is in A; thus $C = 0$. The amplitude of the oscillation is equal to the absolute value of A.

Use the given information to evaluate the acceleration a_0 resulting from the driving force. This information is then used to find the force F_0.

MODULE 10   **Thermodynamics**

The laws of thermodynamics allow us to express the relationship between thermal energy transfer, work, and internal energy. These laws also provide limits on the efficiency of thermal processes.

    The classical view of thermodynamics allows us to analyze the effects of thermal energy on the gross properties of matter. Our study centers around what happens to the system. We use the statistical mechanical approach to give insight into the thermal processes at the molecular level and then relate the properties to measurable quantities such as pressure and temperature.

DEFINITIONS

Internal energy. The collective term for all forms of energy internal to a substance (not influenced by the overall translation or rotation of the body as a whole). Chemical energy, nuclear energy, and thermal energy are internal energy types.

Thermal energy. The total internal mechanical energy of the molecules of a substance.

Thermal equilibrium. The process that occurs when two bodies are placed in thermal contact, and no net heat flows between them. (Heat is another name for thermal energy transfer.)

Temperature. If two bodies placed in thermal contact remain in thermal equilibrium, then they are said to have the same temperature. If they do not, then heat will flow from the body with higher temperature to the one with lower temperature. For most types of systems, temperature is proportional to the average kinetic energy of the molecules.

Heat capacity. A measure of the amount of thermal energy that is required to raise an object's temperature by a specific amount.

Absolute zero. The lowest possible temperature. At this temperature, molecules of a substance have essentially zero thermal energy. Absolute zero corresponds to -273.15 °C.

Zeroth law of thermodynamics. Two objects that are independently in thermal equilibrium with a third object are in thermal equilibrium with each other.

First law of thermodynamics. The first law of thermodynamics is a generalization of the law of conservation of energy to include thermal energy. Any thermal energy absorbed by a system increases the system's internal energy or goes into work done by the system, or both.

Adiabatic process. A process that occurs when there is no net thermal energy transfer.

Isothermal process. A process that occurs when there is no change in temperature.

Second Law of Thermodynamics. In a closed system, the total entropy either increases or stays the same.

Entropy. A measure of the disorder of a system.

USEFUL EQUATIONS   The heat capacity C of a substance is calculated by the equation

$$C = \frac{Q}{\Delta T}$$

where Q is the thermal energy transferred and ΔT is the change in temperature.

The ideal gas law relating pressure, volume, and temperature is described by the equation

$$PV = Nk_{B}T$$

where N is the number of gas molecules in the sample and k_{B} is a constant known as Boltzmann's constant.

The first law of thermodynamics can be expressed in differential form as

$$dQ = dW + dU$$

with W representing the work done by the system and U being the internal energy.

The efficiency of a heat engine is computed by

$$e = \frac{\text{work}}{\text{energy absorbed}} = \frac{Q_h - Q_c}{Q_h}$$

where Q_h is the heat absorbed from the hot reservoir and Q_c is the heat deposited to the cold reservoir.

For a Carnot engine, the efficiency is

$$e = \frac{T_h - T_c}{T_h}$$

with the temperatures being measured on the absolute temperature scale.

The coefficient of performance (COP) of a Carnot heat pump is

$$\text{COP} = \frac{Q_h}{W} = \frac{T_h}{T_h - T_c}$$

The entropy of a system is computed by using the equation

$$dS = \frac{dQ_r}{T}$$

where dQ_r is the change in thermal energy along a reversible path.

Basic Concepts of Thermodynamics

Problem Description
A perfectly insulated calorimeter contains 500 ml of water at 30°C and 25 g of ice at 0°C. Determine the final temperature of the system.

Before we begin...

1. The system is insulated and isolated from its surroundings. It will come to thermal equilibrium. What is meant by thermal equilibrium?

The concept of heat capacity C is discussed on the CD-ROM. Associated with the concept is another concept called specific heat c. Specific heat is the amount of heat per unit mass required to change the temperature of a substance by a given amount.

$$c = \frac{Q}{m\Delta T}$$

For a substance to change phase, an additional heat transfer is required. L_f is the amount of heat transferred per unit mass as the substance melts or freezes, and L_v is the heat transferred per unit mass as it vaporizes or condenses.

$$Q = \pm\, mL_f \qquad \text{or} \qquad Q = \pm\, mL_v$$

2. Identify the given information:

volume of water $V_w =$

initial temperature of water $T_{1w} =$

mass of ice $m_{ice} =$

initial temperature of ice $T_{1\text{-ice}} =$

(*continued on next page . . .*)

Problems

The constant values required for this problem are

specific heat of water $c_w = 4186$ J/kg C°

heat of freezing for water/ice $L_{f\text{-}ice} = 333$ kJ/kg

melting point of ice $T_{mp\text{-}ice} = 0°C$

The density of water is taken to be 1 g/cm³.

Solving the problem

3. We must pay attention to the units in the given information and convert to consistent units throughout the problem. Determine the mass, in kilograms, of the water, using 1.00 ml = 1.00 cm³.

4. In this problem, the net heat exchange will equal zero and the system will reach thermal equilibrium. The water will lose thermal energy, and the ice will use thermal energy to undergo a change of phase. If additional thermal energy is available after all the ice is melted, the ice water will be warmed. Calculate the thermal energy that would be released by lowering the water to the freezing point 0° C.

5. Determine the thermal energy required to melt all of the ice.

6. Solve for the final temperature in this case by applying the law of conservation of energy.

7. If the available energy does not equal or exceed the required energy, not all of the ice will melt. The final temperature will be 0° C. Calculate the amount of the ice that can be melted.

PROBLEM 16 Relative Motion

Before we begin...

1.

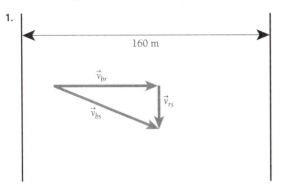

160 m

2. The vector sum is important because the current is pushing the boat downstream at the same time the boat is crossing the river.

Solving the problem

3. The magnitude of the velocity of the boat relative to the shore is

$$v_{bs} = \sqrt{v_{br}^2 + v_{rs}^2} = 2.5 \text{ m/s}$$

4. The time required to cross the river is

$$t = \Delta x/\langle v_x \rangle = (160 \text{ m})/(2.00 \text{ m/s}) = 80.0 \text{ s}$$

5. The distance downstream that the boat will travel is

$$\Delta y = \langle v_y \rangle t = (1.5 \text{ m/s})(80 \text{ s}) = 120 \text{ m}$$

Module 4 Forces

PROBLEM 17 Motion, Newton's First Law, and Force

Before we begin...

1. The net force being exerted on the boat must be zero if it moves at constant velocity.

2.

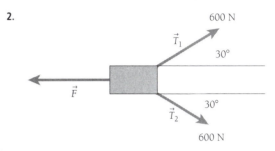

600 N

30°

30°

600 N

Solving the problem

3. $T_{1x} + T_{2x} + F_x = 0$, where $T_{1x} = T_1 \cos 30°$ and $T_{2x} = T_2 \sin 30°$, becomes 520 N + 520 N + F_x = 0; and $T_{1y} + T_{2y} + F_y = 0$, where $T_{1y} = T_1 \sin 30°$ and $T_{2y} = -T_2 \sin 30°$, becomes 300 N − 300 N + F_y = 0.

4. The x and y components of the resistive force $\vec{F}$ are

$$F_x = -1040 \text{ N} \quad \text{and} \quad F_y = 0$$

Therefore $\vec{F} = -1040\hat{i}$ N or 1040 N in the −x direction.

PROBLEM 18 Inertia, Mass, and Weight

Before we begin...

1. The acceleration is equal to the force divided by the mass.

Solving the problem

2. The magnitude of the force exerted by the Earth on the student is

$$\vec{F} = m\vec{g} = (60 \text{ kg})(9.80 \text{ m/s}^2) = 588 \text{ N}$$

toward the center of Earth.

3. Setting $M_E \, \vec{a}_E = 588 \text{ N}$ upward yields

$$\vec{a}_E = (588 \text{ N})/(5.98 \times 10^{24} \text{ kg})$$
$$= 9.83 \times 10^{-23} \text{ m/s}^2 \text{ upward}$$

PROBLEM 19 Newton's Second Law

Before we begin...

1. The given information for this problem is

$\vec{v}_i = 3.0 \text{ m/s}$ $m_1 = 85\text{-kg}$ $\Delta t = 0.5 \text{ s}$
$\vec{v}_f = 4.0 \text{ m/s}$ $m_2 = 58\text{-kg}$

2. The acceleration of the sprinter and the force required to cause this acceleration need to be computed.

Solving the problem

3. The acceleration that the sprinter must experience is

$$\vec{a} = (\vec{v}_f - \vec{v}_i)/t = 2.0 \text{ m/s}^2$$

4. Newton's second law allows you to calculate the net force from the mass and the acceleration.

5. The force on the 85-kg sprinter is

$$\vec{F}_{net} = M\vec{a} = (85 \text{ kg})(2.0 \text{ m/s}^2) = 170 \text{ N}$$

6. This same force would cause the 58-kg sprinter to accelerate

$$\vec{a} = (170 \text{ N})(58 \text{ kg}) = 2.93 \text{ m/s}^2$$

PROBLEM 20 Newton's Third Law

Before we begin...

1.

2. All three blocks must have the same acceleration. The applied force is pushing on the first block.

Solving the problem

3. The net force on each block is related to the block's mass and acceleration as follows:

Block 1 $\vec{F} - \vec{P}_1 = M_1\vec{a}$

Block 2 $\vec{P}_1 - \vec{P}_2 = M_2\vec{a}$

Block 3 $\vec{P}_2 = M_3\vec{a}$

4. Solving the three simultaneous equations for the acceleration yields

$$\vec{a} = 2.00 \text{ m/s}^2$$

5. The contact force acting upon the object is

Block 3 $P_2 = (4.00 \text{ kg})(2.00 \text{ m/s}^2) = 8.00 \text{ N}$

Block 1 $(18.0 \text{ N}) - P_1 = (2.00 \text{ kg})(2.00 \text{ m/s}^2)$
                       $= 4.00 \text{ N}$

$$P_1 = 14.0 \text{ N}$$

As a check, the net force on Block 2 must be 6.00 N. Observe that $P_1 - P_2 = 6.00 \text{ N}$.

PROBLEM 21 Free-Body Diagrams

Before we begin...

1.

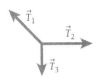

2. The system is in equilibrium; the net force must be zero.

3. The weight of the ball is

$$W = mg = 98.0 \text{ N}$$

Solving the problem

4. The magnitude of the tension T_3 is

$$T_3 = W = 98.0 \text{ N}$$

5. The first free-body diagram shows that the tension $\vec{T}_3$ is pulling vertically downward ($-y$ direction). The directions of $\vec{T}_1$ and $\vec{T}_2$ are

$$T_2 \text{ is at } 0° \text{ and } T_1 \text{ is at } 120°$$

6. Resolving the tensions into their x and y components yields

$$T_{1x} = T_1 \cos 120° = -0.5 \, T_1$$
$$T_{1y} = T_1 \sin 120° = +0.866 \, T_1$$

$$T_{2x} = T_2 \cos 0° = T_2$$
$$T_{2y} = T_2 \sin 0° = 0$$

$$T_{3x} = T_3 \cos 270° = 0$$
$$T_{3y} = T_3 \sin 270° = -98.0 \text{ N}$$

Solving the two equations of equilibrium simultaneously yields

$$T_1 = 113 \text{ N} \quad \text{and} \quad T_2 = 56.5 \text{ N}$$

PROBLEM 22 Free-Body Diagrams

Before we begin...

1. The 4.0-kg object weighs more and so will exert more downward force than the 2.0-kg object. By common sense, the 4.0-kg object will accelerate downward.

2. All of the accelerations must be the same because the strings are assumed not to stretch.

3.

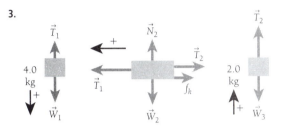

Solving the problem

4. For the three objects, we find

$$\text{4.0 kg} \qquad W_1 - T_1 = M_1 a$$

$$\text{1.0 kg} \qquad T_1 - T_2 = M_2 a$$

$$\text{2.0 kg} \qquad T_2 - W_3 = M_3 a$$

5. Solving the first and third equations for T_1 and T_2, and substituting these results into the second equation, we obtain

$$a = \frac{W_1 - W_3}{M_1 + M_2 + M_3} = \frac{39.2 \text{ N} - 19.6 \text{ N}}{7 \text{ kg}} = 2.8 \text{ m/s}^2$$

6. Substituting this result back into the first and third equations yields

$$T_1 = W_1 - M_1 a$$
$$= 39.2 \text{ N} - (4 \text{ kg})(2.8 \text{ m/s}^2) = 28.0 \text{ N}$$

$$T_2 = W_2 + M_2 a$$
$$= 19.6 \text{ N} + (2 \text{ kg})(2.8 \text{ m/s}^2) = 25.2 \text{ N}$$

As a consistency check, note that this gives a net force on the 1-kg mass of $T_1 - T_2 = 2.8 \text{ N}$. This is just what is needed to give it the predicted acceleration of 2.8 m/s², so our result is consistent.

PROBLEM 23 Centripetal Force

Before we begin...

1.

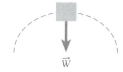

$$\vec{W}$$

2. A normal force does not exist between the driver and the seat because the weight of the driver is providing exactly the centripetal force.

Solving the problem

3. The net force providing the centripetal force is

$$W = mg = F_c = \frac{mv^2}{r}$$

4. The speed of the car is

$$v = \sqrt{rg} = 13.3 \text{ m/s}$$

5. The mass appears in both the net force and the centripetal force. Both sides of the equation can be divided by the mass without altering the relationship.

PROBLEM 24 Fictitious Forces: Motion in Accelerated Reference Frames

Before we begin...

1.

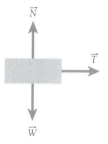

Solving the problem

2. The acceleration of the mass with respect to the inertial frame of reference is

$$\vec{T} = m\vec{a}, \quad \text{so}$$
$$\vec{a} = \frac{\vec{T}}{m} = \frac{18.0 \text{ N}}{5.00 \text{ kg}} = 3.60 \text{ m/s}^2$$

3. When $\vec{a} = 0$, $\vec{T}$ also must be zero.

4. In the noninertial frame of reference, the net force appears to be zero. For this to be accomplished, we must introduce a fictitious force acting to the left and equal in magnitude to the tension $\vec{T}$.

PROBLEM 25 Work

Before we begin...

1. The work done by a variable force is calculated by

$$W = \int_{s_1}^{s_2} \vec{F} \cdot d\vec{s}$$

2. The displacement $d\vec{s}$ is in the positive x direction; $d\vec{s} = dx\hat{i}$.

Solving the problem

3. The integral expression for the work done gives us the following:

$$W = \int_{x=0}^{x=5.0 \text{ m}} (4.0x\hat{i} + 3.0y\hat{j}) \text{ N} \cdot dx\hat{i}$$

Therefore,

$$W = \int_{x=0}^{x=5.0 \text{ m}} (4.0x\hat{i}) \text{ N} \cdot dx\hat{i} + 0$$

or

$$W = (4 \text{ N/m})\frac{x^2}{2} \Big|_0^{5\,\text{m}} = 50.0 \text{ J}$$

Before we begin...

1. The cheerleader has to exert a force equal to the weight of his partner.

$$\vec{F} = -\vec{W} = mg = (50.0 \text{ kg})(9.80 \text{ m/s}^2)$$
$$= 490 \text{ N}\hat{j}$$

2. The displacement is $\Delta \vec{y} = 0.60 \text{ m}\hat{j}$.

Solving the problem

3. The work done for each replication is

$$W = \vec{F} \cdot \vec{s} = Fs \cos \theta$$

Because the force and the displacement are in the same direction, $\cos \theta = 1.00$. For each time the partner was lifted,

$$W = (490 \text{ N})(0.60 \text{ m}) = 294 \text{ J}$$

4. The total work done is $20 \times 294 \text{ J} = 5.88 \times 10^3 \text{ J}$.

PROBLEM 27 Important Examples of Work

Before we begin...

1. The limits of the integration are from $x = 0$ to $x = 0.10$ m

Solving the problem

2. Evaluating the integral between the limits and substituting the value of k into the expression yields

$$W = \int (-kx)dx$$

$$= \frac{-kx^2}{2} = \frac{(10 \text{ N/m})(0.1 \text{ m})^2}{2}$$

$$= -0.5 \text{ J}$$

PROBLEM 28 Work Done to Accelerate a Mass

Before we begin...

1. The units of the coefficients of the three terms are

$$15{,}000 \text{ N} \qquad 10{,}000 \text{ N/m} \qquad 25{,}000 \text{ N/m}^2$$

2. The expression for calculating the work done by a variable force is

$$W = \int_{s_1}^{s_2} \vec{F} \bullet d\vec{s}$$

Solving the problem

3. Substituting the given force into the equation for determining work done by a variable force yields

$$W = \int_{x_1}^{x_2} \vec{F} \bullet d\vec{x} = \int_{0}^{0.6 \text{ m}} (15{,}000 \text{ N} + 10{,}000x \text{ N/m} \\ - 25{,}000x^2 \text{ N/m}^2)dx$$

4. The integral between the limits of $x = 0$ and $x = 0.6$ m is

$$W = 9.00 \times 10^3 \text{ J}$$

5. Repeating between the limits of $x = 0$ and $x = 1.00$ m yields

$$W = 1.17 \times 10^4 \text{ J}$$

The last value is 30 percent greater.

PROBLEM 29 Conservative Forces

Before we begin...

1. Conservative forces and the law of conservation of energy are used to solve this problem.

Solving the problem

2. The potential energy function can be evaluated by integrating the force over the interval of the displacement:

$$U = -\int_0^x (-Ax + Bx^2)dx$$

$$= \frac{Ax^2}{2} - \frac{Bx^3}{3}$$

3. Going from $x = 2.0$ m to $x = 3.0$ m yields

$$\Delta U = \frac{A(3^2 - 2^2)}{2} - \frac{B(3^3 - 2^3)}{3} = \frac{5}{2}A - \frac{19}{3}B$$

Because $\Delta K = -\Delta U$,

$$\Delta K = -\frac{5}{2}A + \frac{19}{3}B$$

PROBLEM 30 Work-Energy Theorem

Before we begin...

1. The given information is

$$m = 15.0 \text{ g} = 0.015 \text{ kg}$$

$$\vec{v}_0 = 0$$

$$\vec{v}_f = 780 \text{ m/s}$$

2. The work done by the net force acting upon a system is equal to the change in kinetic energy of the system.

3. The force accelerates the bullet over $\Delta \vec{x} = 72$ cm $= 0.72$ m.

Solving the problem

4. The change in kinetic energy of the bullet is

$$\Delta K = \frac{1}{2}mv_f^2 - \frac{1}{2}mv_0^2$$
$$= (0.5)(0.015 \text{ kg})(780 \text{ m/s})^2 - 0$$
$$= 4560 \text{ J}$$

5. The definition of work is used to calculate the average force:

$$W = \langle F \rangle \Delta x = 4560 \text{ J}$$

$$\langle F \rangle = 6330 \text{ N}$$

PROBLEM 31 Power

Before we begin...

1.

2. The given information is

$$m = 650 \text{ kg} \qquad v_0 = 0$$
$$v_f = 1.75 \text{ m/s} \qquad \Delta t = 3.00 \text{ s}$$

3. Kinetic energy and gravitational potential energy will be changed by the work done by the elevator motor.

Solving the problem

4. The change in kinetic energy is

$$\Delta K = {}^1\!/_2 m v_f{}^2 - {}^1\!/_2 m v_0{}^2 = 995 \text{ J}$$

5. The height to which the elevator rises in 3.00 s is

$$\Delta y = \langle v \rangle \Delta t = \frac{(0 + 1.75 \text{ m/s})}{2}(3.00 \text{ s}) = 2.63 \text{ m}$$

The change in potential energy is therefore

$$\Delta U = mg\Delta y = (650 \text{ kg})(9.80 \text{ m/s}^2)(2.63 \text{ m})$$
$$= 16{,}750 \text{ J}$$

The total work done by the motor in the first three seconds is 1.77×10^4 J.

6. The average power is calculated as the work done by the motor divided by the time required:

$$P = 5910 \text{ W}$$

7. Once the elevator is moving at a constant speed, the net force must be zero, so the force must equal the weight of the elevator.

Using $P = \vec{F} \cdot \langle \vec{v} \rangle$ after the elevator has reached its constant velocity,

$$P = (650 \text{ kg})(9.80 \text{ m/s}^2)(1.75 \text{ m/s})$$
$$= 1.11 \times 10^4 \text{ W}$$

PROBLEM 32 Conservation of Energy

Before we begin...

1. The law of conservation of energy states that the total mechanical energy of a system (kinetic energy + potential energy) must remain constant in any isolated system of objects that interact only through conservative forces.

2. The given information for the first question is

$$K_i = 30 \text{ J} \qquad \Sigma U_i = 10 \text{ J} \qquad K_f = 18 \text{ J}$$

3. For the second question,

$$\sum U_f = 5 \text{ J}$$

Solving the problem

4. Applying the law of conservation of energy for a conservative system:

$$K_i + \sum U_i = K_f + \sum U_f$$
$$30 \text{ J} + 10 \text{ J} = 18 \text{ J} + \sum U_f$$
$$\sum U_f = 22 \text{ J}$$

5. The total energy E at time t_i is

$$K_i + \sum U_i = 40 \text{ J}$$

6. In the second question, if $\sum U_f = 5$ J, then the total kinetic plus potential energy is no longer 40 J. Work totaling 17 J must have been done by a nonconservative force.

PROBLEM 33 The General Form of Newton's Second Law

Before we begin...

1. The general form of Newton's second law is expressed by the equation

$$\vec{F} = \frac{d\vec{p}}{dt} = m\frac{d\vec{v}}{dt} + \frac{dm}{dt}v_{rel}$$

where $\vec{p}$ is the momentum of the system, $\vec{v}$ is the velocity of the system, and $\vec{v}_{rel}$ is the velocity with which mass is ejected.

2. Thrust is the force exerted on the object by the ejected mass. In this case, the ejected mass is the gas. Thrust is computed by

$$T = v_g \frac{dM_g}{dt}$$

3. The given information is

$$\frac{dM_g}{dt} = 80 \text{ kg/s} \qquad v_g = 2.5 \times 10^3 \text{ m/s}$$

Solving the problem

4. We can evaluate the thrust as follows:

$$T = v_g \frac{dM_g}{dt} = (2.5 \times 10^3 \text{ m/s})(80 \text{ kg/s})$$

$$= 2.0 \times 10^5 \text{ N}$$

Before we begin...

1. The given information is

$$m = 2.0 \text{ kg}$$

$$v_{i\text{(particle initially at rest)}} = 0$$

$$v_{i\text{(particle initially moving)}} = -2.0 \text{ m/s}$$

2. Impulse is related to force and the time interval over which the force is applied by the equation

$$\Delta\vec{p} = \vec{F}\Delta t$$

3. Momentum is defined as $\vec{p} = m\vec{v}$.

Solving the problem

4. The three regions are two triangles and one rectangle with respective areas of

$$4 \text{ N·s, } 4 \text{ N·s and } 4 \text{ N·s}$$

The total impulse therefore is 12 N·s.

5. Final momentum can be expressed in terms of the initial momentum and the impulse as

$$\vec{p}_f = \vec{p}_i + \Delta\vec{p}$$

6. For $\vec{v}_i = 0$, $\vec{p}_f = 0 + 12$ N·s $= 12$ N·s. Since $\vec{p} = m\vec{v}$, $\vec{v}_f = 6.0$ m/s. If $\vec{v}_1 = -2.0$ m/s, $\vec{p}_f = -4$ N·s $+ 12$ N·s $= 8$ N·s. Then, $\vec{v}_f = 4.0$ m/s.

7. A constant force required to give the same value of impulse is

$$12.0 \text{ N·s} = \langle\vec{F}\rangle(5.0 \text{ s})$$

so

$$\langle\vec{F}\rangle = 2.40 \text{ N}$$

Solutions

PROBLEM 35 Perfectly Inelastic Collisions

Before we begin...

1.
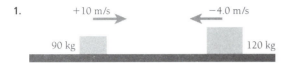

2. The given information is

$m_1 = 90$ kg $\qquad \vec{v}_1 = +10$ m/s

$m_2 = 120$ kg $\qquad \vec{v}_2 = -4.0$ m/s

3. The two masses will stick and move as one.

4. Momentum is always conserved in a collision in an isolated system.

Solving the problem

5. Evaluating the total momentum of the system before the collision, we find

$\vec{p}_T = \vec{p}_1 + \vec{p}_2 = (90$ kg$)(+10$ m/s$)$
$\qquad + (120$ kg$)(-4.0$ m/s$)$
$\qquad = +420$ N·s.

6. Because the momentum after the collision is the same as before the collision, the velocity of the players (who stick together and move as one mass) can be evaluated as

$\vec{p} = (m_1 + m_2)\vec{v}_c$

$\vec{v}_c = (+420$ N·s$)/(210$ kg$) = +2.0$ m/s

Because the + direction was selected as the direction of the halfback's velocity (north) and because the final velocity is also positive, the players are moving north after the collision.

7. Before the collision, the total kinetic energy was

$K_1 + K_2 = \frac{1}{2}m_1v_1^2 + \frac{1}{2}m_2v_2^2 = 5460$ J

After the collision, it becomes

$K_C = \frac{1}{2}(m_1 + m_2)v_c^2 = 420$ J

The work that the players did on each other during the collision converted much of their kinetic energy into other forms. Most of this energy ends up as heat, while smaller amounts end up as sound waves and as lingering vibrations in the players' helmets or skeletal systems.

PROBLEM 36 Perfectly Inelastic Collisions

Before we begin...

1.

2. The given information is

$m_1 = 5$ g $\qquad \vec{v}_1 = (250$ m/s, $20°)$

$m_2 = 3$ g $\qquad \vec{v}_2 = (280$ m/s, $165°)$

3. The two bullets undergo a perfectly inelastic collision.

4. The problem must be solved in two dimensions because the motion is in two dimensions.

Solving the problem

5. The x and y components of momentum for each bullet before the collision are

$p_{1x} = m_1v_1 \cos \theta_1 = 1175$ g·m/s

$p_{1y} = m_1v_1 \sin \theta_1 = 428$ g·m/s

$p_{2x} = m_2v_2 \cos \theta_2 = -811$ g·m/s

$p_{2y} = m_2v_2 \sin \theta_2 = 217$ g·m/s

6. Solving for v_x,

$(m_1 + m_2)v_x = p_{1x} = p_{2x}$

$v_x = \dfrac{(p_{1x} + p_{2x})}{(m_1 + m_2)} = \dfrac{364 \text{ g·m/s}}{8 \text{ g}} = 45.5$ m/s

7. For the y component we find

$v_y = \dfrac{(p_{1x} + p_{2x})}{(m_1 + m_2)} = \dfrac{645 \text{ g·m/s}}{8 \text{ g}} = 80.6$ m/s

8. Evaluating to find the velocity of the combined mass after the collision, we find

$v_c = \sqrt{v_x^2 + v_y^2} = 92.5$ m/s

$\theta = \tan^{-1}\left(\dfrac{v_y}{v_x}\right) = 60.6°$

Saunders Core Concepts in Physics Workbook

Perfectly Elastic Collisions

Before we begin...

1. The given information is

$$m_1 = 2.00 \text{ kg} \quad m_2 = 4.00 \text{ kg} \quad \Delta h = -5 \text{ m}$$

$$v_1 = 0 \qquad\qquad v_2 = 0$$

2. Potential energy is converted into kinetic energy.

3. Linear momentum and kinetic energy are both conserved in an elastic collision.

Solving the problem

4. Using the law of conservation of energy to solve for the speeds of the two objects immediately before the collision,

$$\Delta U + \Delta K = 0 \text{ for each object}$$

(No nonconservative forces are present.)

$$mg\Delta h + (\tfrac{1}{2}mv^2 - 0) = 0$$

The mass cancels in the equation, so solving for the speed of each object gives

$$v = \sqrt{2gh} = 9.90 \text{ m/s}$$

From this, we conclude $v_1 = +9.90$ m/s and $v_2 = -9.90$ m/s immediately before the collision.

5. The total linear momentum and kinetic energy are

$$p_T = (2.00 \text{ kg})(+9.90 \text{ m/s})$$
$$+ (4.00 \text{ kg})(-9.90 \text{ m/s}) = -19.8 \text{ N} \cdot \text{s}$$

$$K_T = \tfrac{1}{2} (2.00 \text{ kg})(9.90 \text{ m/s})^2$$
$$+ \tfrac{1}{2} (4.00 \text{ kg})(9.90 \text{ m/s})^2 = 294 \text{ J}$$

hence, after the collision,

$$2v_1 + 4v_2 = -19.8 \qquad \text{and}$$

$$\tfrac{1}{2}(2)v_1{}^2 + \tfrac{1}{2}(4)v_2{}^2 = 294$$

(The units have been dropped, but will yield velocities in m/s.)

Solving for v_1 yields two possible solutions:

$$v_1 = +9.90 \text{ m/s} \qquad \text{and} \qquad v_1 = -16.5 \text{ m/s}$$

The first solution is rejected; it describes the condition before the collision. The appropriate solution is $v_1 = -16.5$ m/s.

Substitution back into the momentum equation with the value of v_1 yields

$$v_2 = +3.3 \text{ m/s}$$

6. The height to which each block will rise is

$$(0 - \tfrac{1}{2}m_1v_1{}^2) + (m_1gh_1 - 0) = 0$$

When we substitute the speed of the object after the collision, the height is

$$h_1 = \frac{v_1{}^2}{2g} = \frac{(-16.9 \text{ m/s})^2}{2(9.8 \text{ m/s}^2)} = 13.9 \text{ m}$$

Likewise, for object m_2, we obtain

$$h_2 = 0.56 \text{ m}$$

PROBLEM 38 Center of Mass

Before we begin...

1. Equal mass is located above and below the x axis at the same distance from the axis.

2. Based on symmetry, the center of mass will lie on the x axis; $y_{CM} = 0$.

Solving the problem

3. Resolving the distance of the hydrogen atoms to the oxygen atom, we find the x component

$$x = L \cos \theta = (0.100 \text{ nm}) \cos 53°$$
$$= 0.0602 \text{ nm}$$

4. Using the equation for the location of the x_{CM} for a set of discrete particles, we find

$$x_{CM} = \frac{\sum m_i x_i}{\sum m_i}$$

$$= \frac{(15.99 \text{ u})(0)}{+ (1.008 \text{ u})(0.0602 \text{ nm})}$$
$$\frac{+ (1.008 \text{ u})(0.0602 \text{ nm})}{(15.999 + 1.008 + 1.008) \text{ u}}$$

$$= 0.00673 \text{ nm}$$

The coordinates of the center of mass are (0.00673, 0) nm from the oxygen nucleus.

PROBLEM 39 Motion of a System of Particles

Before we begin...

1. The given information is

$$m_1 = 2.0 \text{ kg} \qquad \vec{v}_1 = (2.0\hat{i} - 10t\hat{j}) \text{ m/s}$$
$$m_2 = 3.0 \text{ kg} \qquad \vec{v}_2 = 4.0\hat{i} \text{ m/s}$$

2. The linear momentum of the center of mass of a system is equal to the total linear momentum of the system.

3. Acceleration is the time rate of change of velocity.

Solving the problem

4. From the definition of momentum of the center of mass, we have

$$\vec{p}_{CM} = M\vec{v}_{CM} = m_1\vec{v}_1 + m_2\vec{v}_2$$

$$= (2.0 \text{ kg})[(2.0\hat{i} - 10t\hat{j}) \text{ m/s}]$$
$$+ (3.0 \text{ kg})[(2.0\hat{i} - 10t\hat{j}) \text{ m/s}]$$

$$= (16\hat{i} - 20t\hat{j}) \text{ N·s}$$

5. Solving for the velocity of the center of mass, we find

$$\vec{v}_{CM} = \frac{(16\hat{i} - 20t\hat{j}) \text{ N·s}}{(2.0 \text{ kg} + 3.0 \text{ kg})} = (3.2\hat{i} - 4.0t\hat{j}) \text{ m/s}$$

6. At $t = 0.5$ s, $v_{CM} = (3.2\hat{i} - 2.0\hat{j})$ m/s.

7. The acceleration of the center of mass is

$$\vec{a}_{CM} = \frac{d\vec{v}_{CM}}{dt} = -4.0\hat{j} \text{ m/s}^2$$

8. The total momentum of the system is

$$\vec{p}_{tot} = \vec{p}_{CM} = M\vec{v}_{CM}$$

$$= 5.0 \text{ kg} [(3.21\hat{i} - 2.0\hat{j}) \text{ m/s}]$$

$$= (16\hat{i} - 10\hat{j}) \text{ N·s}$$

Motion of a System of Particles

Before we begin...

1. The given information is

$M = 8.0$ kg $m = 2.0$ kg

$L = 6$ m $h = 2$ m

2. Because there is no external force with an x component acting on the system, x_{CM} will not move. This tells us that the momentum of the center of mass $\Delta p_{CM(x)}$ of the system was and will remain at rest.

Solving the problem

3. The expression for dm in terms of the surface mass density, M/A, and the area of the strip ydx is

$dm = (M/A)ydx$

4. Evaluating the x_{CM}, we obtain

$$x_{CM} = \frac{1}{M}\int x \frac{M}{A} ydx$$

From similar triangles $y/x = h/L$, so that $y = (h/L)x$. Also recall that $A = \frac{1}{2} Lh$.

Substituting into the expression for the x_{CM}, we get

$$x_{CM} = \frac{1}{M}\int x \frac{M}{1/2hL}(h/L)xdx = \frac{2}{L^2}\int x^2 dx = \frac{2L}{3}$$

5. Computing the location of the center of mass of the total system by placing M at the x_{CM} of the triangle and m at the position $x = L$ yields

$$x_{CM} = \frac{(8.0 \text{ kg})(4 \text{ m}) + (2.0 \text{ kg})(6 \text{ m})}{8 \text{ kg} + 2 \text{ kg}} = 4.4 \text{ m}$$

6. Setting the value calculated equal to the situation where mass m is located at d and M is located at $(x_{CM \text{ triangle}} + d)$ gives

$$\frac{(8.0 \text{ kg})(4 \text{ m} + d) + (2.0 \text{ kg})d}{(8 \text{ kg} + 2 \text{ kg})} = 4.4 \text{ m}$$

Solving for d, we find

$d = 1.20$ m

Basic Concepts in Rotational Kinematics

Before we begin...

1. The given information is

$v_i = 0$ $v_f = 25$ m/s $\theta = 1.25$ rev $R = 1.00$ m

2. Angular speed is related to tangential or linear speed by the equation

$$\omega = \frac{v}{R}$$

Solving the problem

3. Converting the initial and final speeds to angular speeds using the linear transformation equations yields

$\omega_i = 0$ $\omega_f = 25.0$ rad/s

4. An equation from rotational kinematics that involves initial and final angular speeds, the angle through which the acceleration occurs, and the angular acceleration is

$2\alpha\theta = \omega_f^2 - \omega_i^2$

5. Converting the angle from revolutions to radians gives

$\theta = (1.25 \text{ rev})(2\pi \text{ rad/rev}) = 7.85 \text{ rad}$

6. Solving for angular acceleration we find

$$\alpha = \frac{\omega_f^2 - \omega_i^2}{2\theta} = \frac{(25.0 \text{ rad/s})^2}{2(7.85 \text{ rad})} = 39.8 \text{ rad/s}^2$$

7. To solve for the time, use

$$\alpha = \frac{\Delta\omega}{\Delta t}$$

$$\Delta t = \frac{\Delta\omega}{\alpha} = \frac{25.0 \text{ rad/s}}{39.8 \text{ rad/s}^2} = 0.628 \text{ s}$$

Solutions

PROBLEM 42 Rotational Energy

Before we begin...

1. The given information is

$$m = 215 \text{ g} = 0.215 \text{ kg} \qquad r = d/2 = 0.0319 \text{ m}$$

$$\Delta x = 3.00 \text{m} \qquad\qquad \theta = 25°$$

$$\Delta t = 1.50 \text{ s} \qquad\qquad v_i = 0$$

2. The length of the cylinder does not matter as long as it is rolling on its side.

3. The total energy of the system at the top of the incline will be equal to the total energy when the cylinder reaches the bottom.

Solving the problem

4. Computing ω_f, we find

$$\langle v \rangle = \Delta x / \Delta t \qquad \text{and} \qquad \langle v \rangle = {}^1\!/_2(v_i + v_f)$$

so

$$v_i + v_f = 2\Delta x / \Delta t$$

$$v_f = 4.00 \text{ m/s}$$

from which

$$\omega_f = v_f / r = 125 \text{ rad/s}$$

5. The initial gravitational potential energy is

$$U_i = mgh$$
$$= (0.215 \text{ kg})(9.80 \text{ m/s}^2)[(3 \text{ m}) \sin 25°]$$

6. Setting the total energy at the top of the incline equal to the total energy at the bottom, we obtain

$$(K_t + K_r + U)_{\text{top}} = (K_t + K_r + U)_{\text{bottom}}$$

$$0 + 0 + mgh = {}^1\!/_2 mv^2 + {}^1\!/_2 I\omega^2 + 0$$

We can then solve for I:

$$I = \frac{m(2gh + v^2)}{\omega^2} = 1.21 \times 10^{-4} \text{ kg·m}^2$$

PROBLEM 43 Moment of Inertia of Rigid Bodies

Before we begin...

1. The moment of inertia of a point mass rotating about a fixed axis is computed by $I = mr^2$, where r is the distance from the point of rotation to the mass m.

Solving the problem

2. The equation for the total moment of inertia of the two-object system is

$$I = Mx^2 + m(L - x)^2$$

3. Taking the first derivative of the function and setting it equal to zero to find the following values of x that satisfy the equation,

$$\frac{dI}{dx} = 2Mx - 2m(L - x) = 0$$

$$x = \frac{mL}{m + M}$$

4. By determining that

$$\frac{d^2 I}{dx^2} = 2m + 2M$$

is positive, we see that the value of x is a minimum.

5. The center of mass of the system can be calculated according to the methods of Module 6, Linear Momentum, and shown to be the same as the value of x in the solution for question 3.

$$x_{\text{CM}} = \frac{\sum x_i m_i}{\sum m_i} = \frac{mL + M(0)}{(m + M)} = \frac{mL}{(m + M)}$$

Calculating I_{CM}, we find

$$I_{\text{CM}} = M\left(\frac{mL}{m + M}\right)^2 + m\left(L - \frac{mL}{m + M}\right)^2$$

Simplifying this expression yields

$$I_{\text{CM}} = \left(\frac{Mm}{m + M}\right)L^2 = \mu L^2$$

Before we begin...

1. The given information in unit vector notation is

$$\vec{r} = (\hat{i} + 3\hat{j}) \text{ m}$$

$$\vec{F} = (3\hat{i} + 2\hat{j}) \text{ N}$$

2.
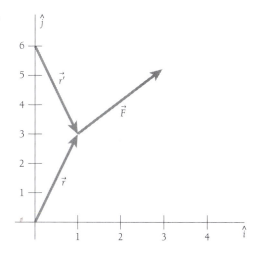

3. Torque is related to position and applied force by

$$\tau = \vec{r} \times \vec{F}$$

Solving the problem

4. We can use the determinant form to write the expression for $\tau = \vec{r} \times \vec{F}$ and solve for the torque:

$$\vec{r} \times \vec{F} = \begin{vmatrix} \hat{i} & \hat{j} & \hat{k} \\ 1 & 3 & 0 \\ 3 & 2 & 0 \end{vmatrix} = (2 - 9)\hat{k} = -7\hat{k} \text{ N·m}$$

5. The position vector in terms of the new point of rotation is

$$\vec{r}' = (1, 3, 0) - (0, 6, 0) = (1, -3, 0) \quad \text{or}$$

$$\vec{r}' = (\hat{i} - 3\hat{j}) \text{ m}$$

6. Solving for the torque about the new point of rotation yields

$$\vec{r}' \times \vec{F} = \begin{vmatrix} \hat{i} & \hat{j} & \hat{k} \\ 1 & -3 & 0 \\ 3 & 2 & 0 \end{vmatrix} = (2 + 9)\hat{k} = 11\hat{k} \text{ N·m}$$

7. It should be noted that $\vec{r}$ and $\vec{F}$ are both in the xy plane. The torque vectors are perpendicular to the r, F plane. This places them on either the z or $-z$ axis.

Solutions

PROBLEM 45 Work and Energy Problem

Before we begin...

1. The given information is

$$M = 100 \text{ kg} \qquad R = 0.5 \text{ m} \qquad F_n = 70 \text{ N}$$

$$\Delta t = 6.0 \text{ s} \qquad \omega_i = 50 \text{ rev/min}$$

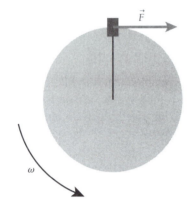

2. The frictional torque opposes the direction of rotation and thus decreases the rotational speed of the wheel.

3. The final rotational kinetic energy of the wheel will be zero.

Solving the problem

4. Using the equation $I = \frac{1}{2}MR^2$ for a solid disk, we calculate the moment of inertia of the wheel to be

$$I = \frac{1}{2}(100 \text{ kg})(0.5 \text{ m})^2 = 12.5 \text{ kg·m}^2$$

5. The angular speed is converted to

$$\omega_i = (50 \text{ rev/min})(2 \pi \text{ rad/rev})(1 \text{ min/60 s})$$
$$= 5.23 \text{ rad/s}$$

This must be done so that the expression will be in fundamental units.

6. The initial rotational kinetic energy and the change in kinetic energy during the problem are

$$K_i = \frac{1}{2}I\omega_i^2 = 171 \text{ J}$$

$$\Delta K = (0 - 171 \text{ J}) = -171 \text{ J}$$

7. The formula for the work done by a constant net torque is

$$\tau_{net}\Delta\theta = \Delta K$$

8. To find $\Delta\theta$,

$$(\omega_i + \omega_f)/2 = \Delta\theta\Delta/t \qquad \text{so} \qquad \Delta\theta = 15.7 \text{ rad}$$

9. Using the work-energy theorem to solve for the torque,

$$(\tau_{net})(15.7 \text{ rad}) = 171 \text{ J}$$

$$\tau_{net} = 10.9 \text{ N·m}$$

(Note that the negative sign indicates the direction of the torque and is therefore ignored in computing the magnitude.)

PROBLEM 46 Rolling Motion

Before we begin...

1.

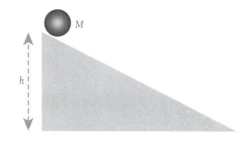

2. The moments of inertia for the objects are

$$I_{\text{disk}} = \frac{1}{2}MR^2 \qquad I_{\text{hoop}} = MR^2$$

3. Each object has potential energy at the top of the incline.

Solving the problem

4. A rolling object has both rotational kinetic energy and translational kinetic energy. The expressions for the type of kinetic energy are

$$K_{\text{rot}} = \frac{1}{2}I\omega^2 \qquad K_{\text{tran}} = \frac{1}{2}Mv^2$$

5. The law of conservation of energy as it applies to this problem is

$$U_i + K_{i\text{-rot}} + K_{i\text{-tran}} = U_f + K_{f\text{-rot}} + K_{f\text{-tran}}$$

Substituting the known quantities into this general equation,

$$Mgh + 0 + 0 = 0 + \frac{1}{2}I\omega_f{}^2 + \frac{1}{2}Mv_f{}^2$$

For a disk,

$$Mgh + 0 + 0 = 0 + \frac{1}{2}\left(\frac{1}{2}MR^2\right)\omega_f{}^2 + \frac{1}{2}Mv_f{}^2$$

For a hoop,

$$Mgh + 0 + 0 = 0 + \frac{1}{2}(MR^2)\omega_f{}^2 + \frac{1}{2}Mv_f{}^2$$

Using $v = R\omega$, we can solve after substitution

For a disk,

$$v = \sqrt{\frac{4}{3}gh}$$

For a hoop,

$$v = \sqrt{gh}$$

6. The disk has the greater speed at the bottom.

Because the objects cover the same distance, the one with the greater average speed will reach the bottom first. Because they have the same initial speed, the one with the greater final speed will have the greater average speed as well.

7. The disk stores less of its kinetic energy in rotation, leaving more for translational kinetic energy.

Before we begin...

1. The angular momentum of a particle rotating in a constant circular path is computed by $L = \vec{R} \times \vec{p}$ where $\vec{p} = m\vec{v}$. $L = Rmv \sin \theta$, where θ is the angle between $\vec{R}$ and $\vec{p}$.

2.

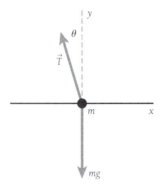

3. A particle moving in a constant circular path must be subjected to a net centripetal force.

Solving the problem

4. The tension $\vec{T}$ is resolved into its x and y components:

$$T_x = T \sin \theta \quad \text{and} \quad T_y = T \cos \theta$$

For the x forces,

$$T \sin \theta = ma = mv^2/R$$

In the y direction,

$$T \cos \theta - mg = 0$$

5. Solving these equations for T and equating the expressions,

$$\frac{mg}{\cos \theta} = \frac{mv^2}{R \sin \theta}$$

so,

$$v = \sqrt{\frac{Rg \sin \theta}{\cos \theta}}$$

6. Angular momentum is expressed as $L = Rmv \sin 90°$, because $\vec{R}$ and $\vec{p}$ are tangential to each other. Thus,

$$L = Rm\sqrt{\frac{Rg \sin \theta}{\cos \theta}} = \sqrt{\frac{m^2gR^3 \sin \theta}{\cos \theta}}$$

Because $R = l \sin \theta$, then

$$L = \sqrt{\frac{m^2gl^3 \sin^4 \theta}{\cos \theta}}$$

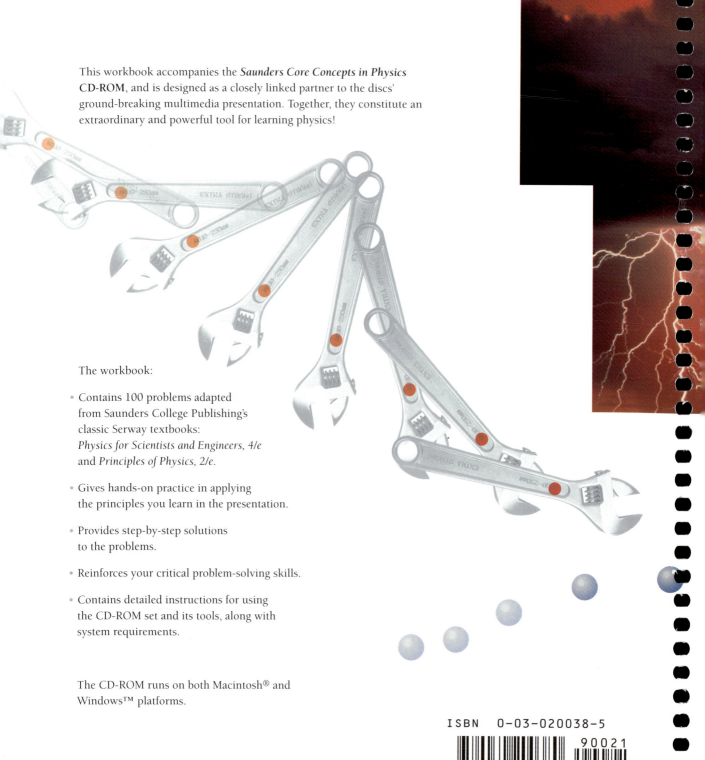

This workbook accompanies the *Saunders Core Concepts in Physics*
CD-ROM, and is designed as a closely linked partner to the discs'
ground-breaking multimedia presentation. Together, they constitute an
extraordinary and powerful tool for learning physics!

The workbook:

- Contains 100 problems adapted
  from Saunders College Publishing's
  classic Serway textbooks:
  Physics for Scientists and Engineers, 4/e
  and *Principles of Physics, 2/e.*

- Gives hands-on practice in applying
  the principles you learn in the presentation.

- Provides step-by-step solutions
  to the problems.

- Reinforces your critical problem-solving skills.

- Contains detailed instructions for using
  the CD-ROM set and its tools, along with
  system requirements.

The CD-ROM runs on both Macintosh® and
Windows™ platforms.

ISBN   0-03-020038-5

90021

9 780030 200380

Saunders
College
Publishing

Archipelago